TABLE OF CONTENTS

What Is a Shark? 2
Prehistoric Sharks 3
Meet the Megalodon 4
A Mouth Full of Teeth 6
Top Ten Most Dangerous Sharks 8
   #10 Oceanic Whitetip Shark 10
   #9 Bronze Whaler Shark 12
   #8 Hammerhead Shark 14
   #7 Spinner Shark 16
   #6 Wobbegong Shark 18
   #5 Sand Tiger Shark 20
   #4 Blacktip Shark 22
   #3 Bull Shark 24
   #2 Tiger Shark 26
   #1 Great White Shark 28
More Cool Sharks
   Whale Shark 30
   Megamouth Shark 32
   Blind Shark 34
   Broadnose Sevengill Shark 36
   Angel Shark 38
   California Horn Shark 40
   Velvet Belly Lantern Shark 42
   Crocodile Shark 44
   Lemon Shark 46
Glossary 48

WHAT IS A SHARK?

Sharks are fish. They live in oceans all over the world and even some rivers and lakes. Sharks do not have bones. Their skeleton is made of cartilage.
There are about 500 different kinds of sharks today. All sharks are carnivores, meaning they eat meat. Most sharks swim even when they are sleeping! If they stop moving, they will stop breathing.

SHARK BITE

A group of sharks is called a school or shoal.

PREHISTORIC SHARKS

Sharks have been around for more than 350 million years—even 100 million years before the dinosaurs. That makes them **prehistoric**!

Modern sharks showed up during the **Jurassic period** about 150 million years ago. The most famous prehistoric shark of all is the Megalodon! The Megalodon (pronounced MEG-uh-luh-don) is said to be the largest marine **predator** to have ever lived. Its name means "giant tooth" and it's pretty obvious why. The largest recorded tooth found measures 7.5 inches (19.1 cm)!

7.5 INCHES (19.1 CM)

The Megalodon weighed an estimated 75 tons (68 tonnes) or more. The only marine animal larger is the modern-day blue whale.

MEET THE MEGALODON

From the tip of the nose to the end of its tail, scientists believe the Megalodon may have grown as long as 70 feet (21.3 m) long!

Megalodon had a huge appetite! It is believed they often ate their favorite prey, large whales, by tearing off their fins first.

According to estimates, the Megalodon had a stronger bite than even the most terrifying dinosaur—the Tyrannosaurus rex!

Today, Megalodon teeth can be found all over the world. This is because they lived in every ocean.

The Megalodon's bite was 11 feet (3.4 m) wide and almost 9 feet (2.7 m) tall. That's big enough to swallow a whole school bus!

A Megalodon tooth could grow to more than 7 inches (18 cm) long. A great white shark's tooth is only 3 inches (7.6 cm)!

A MOUTH FULL OF TEETH

Sharks may have up to 3,000 teeth. They are arranged in rows. Most sharks have five rows of teeth. Sharks use their teeth to bite large chunks out of their prey, gulping down whole pieces at a time.

Megalodon had five rows of more than 270 sharp teeth.

Great white sharks have 300 super-sharp triangular teeth arranged in rows.

The bull shark can have up to 50 rows of teeth in its mouth!

A sand tiger shark's row of sharp, pointed teeth can be seen even when the mouth is closed (1 inch, 2.5 cm).

Lemon sharks have curved teeth so they can hang on to slippery fish (.75 inch, 19 mm).

Shortfin mako sharks have long, sharp teeth that stick out of their mouth (1.2 inches, 3 cm).

Tiger shark teeth aren't just sharp. They have saw-like edges to cut through shells (1 inch wide by 1 inch tall, 2.5 cm by 2.5 cm).

The longnose sawshark has a long, toothy snout, as well as sharp teeth in its mouth (.3 inch, 7.6 mm).

Hammerhead shark teeth are small and smooth (.25 to .75 inch, 6.4 mm to 19 mm).

SHARK BITE

Sharks are constantly losing their teeth, but that's okay because new ones grow in their place!

TOP TEN MOST DANGEROUS SHARKS

Sharks do not normally attack humans. Most of the time, the shark thinks a person looks like its favorite food, such as a seal or sea lion. Sharks attack fewer than 100 people each year. You are more likely to be stung by a bee or get struck by lightning.

10 Oceanic Whitetip Shark (Length: Up to 9.8 feet or 3 m)
The oceanic whitetip is a quick, bold, and **aggressive** hunter. It has been known to attack shipwreck survivors.

9 Bronze Whaler Shark (Length: Up to 11 feet or 3.5 m)
Also known as the copper shark, this shark often hunts in large groups.

8 Hammerhead Shark (Length: Up to 20 feet or 6.1 m)
Hammerheads hunt at night and take sudden and sharp turns.

7 Spinner Shark (Length: Up to 9.1 feet or 2.8 m)
Spinner sharks are fast and will jump out of the water to catch their prey.

6. Wobbegong Shark (Length: Up to 10 feet or 3 m)
Wobbegong sharks spend most of their time on the ocean floor.

5. Sand Tiger Shark (Length: Up to 10.5 feet or 3.2 m)
Sand tiger sharks troll the ocean floor very close to shore.

4. Blacktip Shark (Length: Up to 5.2 feet or 1.6 m)
During the breeding season, more than 10,000 blacktip sharks hang out along the Florida coastline.

3. Bull Shark (Length: Up to 11 feet or 3.5 m)
Some scientists believe that some shark bites said to be from great whites are actually from bull sharks.

2. Tiger Shark (Length: Up to 14 feet or 4.2 m)
Tiger sharks will eat almost anything!

1. Great White Shark (Length: Up to 20 feet or 6.1 m)
Unfortunately for surfers, great whites are curious and like to "taste test" objects that look interesting.

#10 OCEANIC WHITETIP SHARK

- Oceanic whitetips have traveling companions including dolphinfish, remoras, and pilotfish. They eat parasites off the shark and even clean their teeth.
- It is difficult to find food in the open ocean, so oceanic whitetips may eat only once a month.
- Oceanic whitetip sharks get their name from the white tips on their fins.

SHARK BITE

These lone hunters are sometimes found with female pilot whales, because the whales know where to find squid.

SHARK FAST FACTS

Length: up to 13 feet (4 m)
Weight: 370 pounds (168 kg)
Food: squid, fish, seabirds, sea turtles, and carrion
Swim speed: average 5 mph (8 kph), but they are capable of quick bursts of speed
Location: warm waters worldwide; open ocean, rocky reefs, and coral reefs

#9 BRONZE WHALER SHARK

SHARK BITE

The bronze whaler can be very bold when food is around.

12

- Bronze whaler sharks have a highly recognizable ridge right between the dorsal fins that is bronze in color.

- Bronze whalers migrate every spring and fall, traveling up to 820 miles (1,320 km).

- These sharks have narrow, hook-shaped teeth.

- Bronze whalers hunt in large groups.

SHARK FAST FACTS

Length: 11 feet (3.4 m)
Weight: up to 672 pounds (305 kg)
Food: fish, octopuses, smaller sharks, and rays
Swim speed: fast; exact speed unknown
Location: temperate waters around the world

#8 HAMMERHEAD SHARK

- Hammerhead sharks use their heads to trap their favorite meal, stingrays, on the ocean floor.
- Usually, hammerheads like to be alone, but some can be found in schools of 100 or more.
- Female hammerheads can give birth to up to 50 pups at one time.

SHARK BITE

There are nine different kinds of hammerhead sharks:
- winghead
- scalloped bonnethead
- whitefin hammerhead
- scalloped hammerhead
- scoophead
- great hammerhead
- bonnethead
- smalleye hammerhead
- smooth hammerhead

SHARK FAST FACTS

Length: up to 13 feet (4 m)
Weight: up to 500 pounds (226.8 kg)
Food: stingrays, bony fish, crabs, squid, lobster, and other sea creatures
Swim speed: 25 mph (40 kph)
Location: oceans all over the world

#7 SPINNER SHARK

SHARK BITE

This shark is often mistaken for a blacktip shark because its fins have gray or black tips.

- Spinner sharks leap out of the water and can spin up to three times in the air before falling back in the water.
- They hunt in groups and swim up through schools of fish with their mouths open wide, all while spinning.
- Spinner sharks have narrow, triangular teeth.
- This shark is often hunted and sold as a delicacy.

SHARK FAST FACTS

Length: up to 9.1 feet (2.8 m)
Weight: 120–200 pounds (54.4–90.7 kg)
Food: fish, stingrays, octopuses, and squid
Swim speed: up to 46 mph (74 kph) when leaping
Location: shallow coastal waters in parts of the Atlantic, Pacific, and Indian Oceans, and the Mediterranean Sea

#6 Wobbegong Shark

18

SHARK BITE

Wobbegong comes from a language of the First Nations People of Australia and means "shaggy beard."

- Wobbegong sharks are known as carpet sharks because they live on the ocean floor and have patterns on their skin.
- This group includes 12 species of carpet sharks such as the spotted wobbegong, the floral banded wobbegong, and the tasselled wobbegong.
- Wobbegongs sleep during the day and hunt at night.
- They wait for prey to come close to them and then attack.

SHARK FAST FACTS

Length: 6–10 feet (1.8–3 m)
Weight: 154 pounds (70 kg)
Food: fish and invertebrates
Swim speed: slow, mostly sits on the ocean floor
Location: coral reefs in Asia, New Guinea, and Australia

#5 SAND TIGER SHARK

SHARK BITE

Sand tiger sharks and tiger sharks are different kinds of sharks.

- Sand tiger sharks can come to the surface to gulp air to help them float and watch for prey without moving in the water.
- They always look as if they are staring, because they do not have eyelids.
- To clean their eyes, they roll them back in their sockets.
- Sand tiger sharks can pump water over their gills so they can rest on the ocean bottom, unlike some other sharks that have to keep moving in order to breathe, even when they sleep.

SHARK FAST FACTS

Length: 6.5–10 feet (2–3 m)
Weight: up to 350 pounds (159 kg)
Food: herring, snappers, eels, mackerels, other fish, and sometimes other sharks
Swim speed: 12 mph (19 kph)
Location: warm waters close to shore

#4 BLACKTIP SHARK

SHARK BITE

Blacktip sharks live around river mouths, muddy bays, and mangrove swamps, but they do not go into fresh water.

- Blacktip sharks sometimes hunt by leaping out of the water and splashing down on their backs.
- They get their name from their black-tipped fins.
- Blacktips can quickly adjust their eyes in low light.

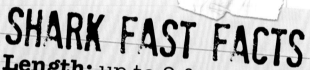

SHARK FAST FACTS
Length: up to 8 feet (2.4 m)
Weight: up to 220 pounds (100 kg)
Food: fish, skates, stingrays, squid, and **crustaceans**
Swim speed: 14 mph (22.5 kph)
Location: warm coastal waters around the world

#3 BULL SHARK

24

- Bull sharks hunt both during the day and at night.
- They will eat almost anything.
- Bull sharks are one of the few sharks that can live and hunt in salt water and fresh water.
- Sometimes bull sharks hunt in groups.

SHARK FAST FACTS

Length: 7–11.5 feet (2.1–3.5 m)
Weight: up to 500 pounds (227 kg)
Food: fish, small sharks, turtles, birds, and dolphins
Swim speed: 25 mph (40 kph)
Location: all over the world, near the shoreline

SHARK BITE

A bull shark has the strongest bite of any shark.

#2 TIGER SHARK

- Tiger sharks are known as the garbage cans of the sea. They will eat anything!
- Female tiger sharks are bigger than male tiger sharks.
- When tiger sharks find prey, they move slowly, stalking it.
- Tiger sharks are the only kind of sharks that create eggs and then give birth to live young.

SHARK BITE

A license plate, plastic bottles, nails, oil cans, and an old tire have been found in the stomachs of tiger sharks.

SHARK FAST FACTS

Length: 10–14 feet (3–4.3 m)
Weight: up to 1,400 pounds (635 kg)
Food: stingrays, sea turtles, clams, sea snakes, seals, birds, and squid
Swim speed: 20 mph (32 kph)
Location: tropical and **subtropical** water all over the world

#1 GREAT WHITE SHARK

- Great white sharks can smell one drop of blood in 25 gallons (96.9 L) of water and can sense even a tiny amount of blood up to 3 miles (5 km) away.

- They have torpedo-shaped bodies that help them swim fast.

- Great whites are blue gray on top. Their bellies are white. This makes it difficult to see them from above and below.

SHARK FAST FACTS
Length: 15–20 feet (4.6–6 m)
Weight: 5,000 pounds (2,268 kg) or more
Food: fish, seals, sea lions, small whales, and sea turtles
Swim speed: 15 mph (24 kph)
Location: cool, coastal waters around the world

SHARK BITE

Great white sharks are almost the size of a school bus.

MORE COOL SHARKS
WHALE SHARK

30

- Whale sharks feed close to the surface of the ocean. They filter plankton through their gills.
- These sharks have 3,000 tiny teeth that are not used to eat.
- Whale sharks are not whales. They are sharks.

SHARK FAST FACTS
Length: 18-40 feet (5.5-12 m)
Weight: average 20.6 tons (18.7 tonnes)
Food: plankton, small fish, and squid
Swim speed: no more than 3 mph (4.8 kph)
Location: warm waters around the equator worldwide

SHARK BITE
The whale shark was thought to develop about 60 million years ago and is one of the oldest species on Earth today.

MEGAMOUTH SHARK

SHARK BITE

The megamouth shark is not a big danger, because it has a weak body and swims poorly.

- Scientists found the first megamouth shark in 1976. Since then, there have only been 63 sightings.
- The megamouth is a giant shark that has a mouth like a filter, eating mostly shrimp and plankton.
- This shark swims in deep waters with its mouth open, but it does breach the surface at night.
- Megamouth sharks have up to 50 rows of teeth in their upper jaw and up to 75 rows of teeth in their lower jaw.

SHARK FAST FACTS

Length: up to 18 feet (5.5 m)
Weight: up to 2,679 pounds (1,215 kg)
Food: shrimp, plankton, small fish, and jellyfish
Swim speed: slow
Location: worldwide but in small patches of ocean

BLIND SHARK

34

- Blind sharks aren't actually blind. They get their name because they close their eyes when threatened.

- Often caught in receding pools of water, the blind shark can survive out of water for up to 18 hours.

SHARK FAST FACTS

Length: 3.9 feet (1.2 m)
Weight: up to 45 pounds (20.4 kg)
Food: sea anemones, squid, crustaceans, and small fish
Swim speed: slow
Location: coast of eastern Australia in rocky areas and seagrass beds

SHARK BITE

The blind shark can pull its eyeballs inward and close them with thick eyelids.

BROADNOSE SEVENGILL SHARK

SHARK BITE

This shark has seven gills—most other sharks have five.

- Sevengill sharks sometimes hunt in packs, working as a team to catch large prey.

- This shark will eat almost anything, including dead animals, so it is both a hunter and a scavenger.

- Its lower jaw has large, comb-shaped teeth which are good for tearing and cutting into prey.

SHARK FAST FACTS

Length: up to 9.8 feet (3 m)
Weight: about 236 pounds (107 kg)
Food: other sharks, bat rays, harbor seals, dolphins, crabs, and the remains of dead animals
Swim speed: unknown, with bursts of speed when hunting
Location: coastal waters less than 164 feet (50 m) deep in all the oceans

ANGEL SHARK

SHARK BITE

The angel shark's nickname is "sand devil," because it hides on the sea floor and strikes with needle-like teeth.

- Angel sharks have flattened bodies and broad pectoral fins, or "wings," that make them look like a stingray.
- They are not fast swimmers, so they ambush their prey by hiding on the sea floor.
- Angel sharks can strike their prey in one-tenth of a second.
- These sharks have spiracles on the top of their head that pump water through the gills while they're hiding in the sand.

SHARK FAST FACTS

Length: 4.9 feet (1.5 m)
Weight: up to 77 pounds (34 kg)
Food: fish, crustaceans, clams, and mussels
Swim speed: 2.5 mph (4 kph)
Location: shallow coastal waters throughout the world

CALIFORNIA HORN SHARK

SHARK BITE

This shark is also known as the walking crab cruncher.

40

- The California horn shark has a small mouth full of small jagged teeth.

- It has two fins up front that it uses to crawl over rocks in shallow water.

- The egg of a California horn shark is a leathery corkscrew shape that the female wedges into a safe space.

- This shark has two stinging spines on its back.

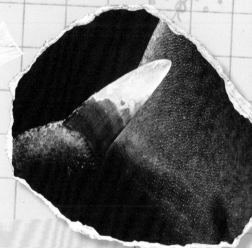

SHARK FAST FACTS
Length: 4 feet (1.2 m)
Weight: up to 60 pounds (27 kg)
Food: snails, crabs, sea urchins, and small fish
Swim speed: clumsy swimmer at 2.3 mph (3.7 kph)
Location: coastal waters worldwide

VELVET BELLY LANTERN SHARK

SHARK BITE

One type of lantern shark is the dwarf lantern shark—hardly ever seen by humans. They are the smallest of all shark species, only 4 inches (10.2 cm) long!

- This shark can make sharp spines on its back light up. One scientist called them *light sabers*!

- The lantern shark was named for a type of lamp used by people!

- Some might use their light to "talk" to each other. The patterns of the light cells might be sending messages to other lantern sharks.

SHARK FAST FACTS

Length: 11–18 inches (28–45.7 cm)
Weight: 2 pounds (.9 kg)
Food: shellfish, small fish, and squid
Swim speed: unknown
Location: east Atlantic, western Mediterranean, the Azores, the Canary Islands, and Cape Verde

CROCODILE SHARK

SHARK BITE

"Crocodile shark" comes from its Japanese name mizuwani, meaning "water crocodile."

- Crocodile sharks have really large eyes that help them hunt in darkness.
- When a crocodile shark has pups, the stronger ones may eat the weaker ones while they're still in their mother's womb.
- When taken out of the water, the crocodile shark will snap its sharp teeth violently.
- The crocodile shark has an oily liver that helps it float.

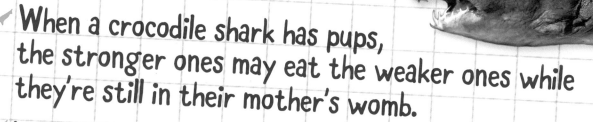

SHARK FAST FACTS

Length: 3.9 feet (1.2 m)
Weight: unknown
Food: fish, squid, and shrimp
Swim speed: swift moving
Location: deep tropical waters worldwide

LEMON SHARK

- Lemon sharks get their name from their yellow coloring.
- These sharks live near the sandy bottom but can be found in fresh water too.

- Lemon sharks cannot see very well. They rely on their sense of smell and the ampullae of Lorenzini to detect prey.
- It is one of the few sharks that does well in captivity.

SHARK FAST FACTS

Length: 8–10 feet (2.4–3 m)
Weight: up to 420 pounds (190.5 kg)
Food: fish
Swim speed: average 2 mph (3.2 kph)
Location: warm shallow waters of Africa, Australia, North America, and South America

SHARK BITE

Lemon sharks are very social with one another.

GLOSSARY

Aggressive — mean and unfriendly in one's actions; very bold and forceful

Ampullae of Lorenzini — any of the pores on the snouts of marine sharks and rays that contain receptors highly sensitive to weak electric fields

Breach — rise or break through the surface of the water

Carnivore — an animal that eats the flesh of other animals

Cartilage — a tough, fibrous substance that is not as stiff as bone

Coastal — near the edge of land

Crustacean — an animal with a hard, jointed shell

Delicacy — something pleasing to eat that is considered rare or luxurious

Filter — to separate out

Fresh water — naturally occurring body of water that is not salty

Gills — the organs used for breathing by fish and other animals that live in the water

Jurassic period — a time in Earth's history 210 million years ago

Migrate — to move from one place to another